Test Method Validation for Medical Devices

I0480284

Emmet Tobin

ISBN-13: 978-1974211579

ISBN-10: 1974211576

Contents

Introduction

Regulatory Overview

Definitions and Key Concepts

ValidaitonResources.org

Introduction

This guidebook covers the design, execution and analysis of test method validation for medical devices. Test method validation involves the formal documentation of a test method used to capture and analyse data or information. The reason test methods need to be validated is to confirm that they are suitable and fit for the intended purpose. Secondly, and of equal importance is the need to verify that the test method performs to an acceptable level and is reliable and trustworthy over time. After all, test methods are used to assess product outputs such as dimensions, material strength and product functions. Getting this wrong will lead nowhere very quickly, so it is important to have confidence in the results of testing.

Validation studies must demonstrate method capabilities in the testing environment. As a result, validation studies allow the formal documentation of the ruggedness of the test method in real-use conditions (i.e. demonstrating that the precision and accuracy limits are met with different technicians, different production batches and variable test equipment, etc.).

Examples of Test Method Validations

Example 1

A packaging company has a seal strength on the lid of a package. It wants to put in place a test method to test the seal strength of the package. This scenario would call for a test method validation.

Example 2

A medical device incorporates the use of a spring that is used to actuate a valve. The manufacturer of the device wants to develop a test method that examines the tensile strength of the spring on an ongoing basis. This scenario would also call for a test method validation.

Example 3

A contact lens manufacturer uses an optical comparator to measure the diameter of contact lenses during manufacture. The manufacturer must develop and validate a test method to facilitate the measuring of contact lenses.

What is test method validation?

Test method validation is the documented process of ensuring a test method is suitable for its intended use.

The intended use of any system is normally documented and described in a User Requirements Specification. Test method validation involves establishing the performance characteristics and limitations of a method and the identification of influences which may change those characteristics.

Why should TMV be performed?

TMV is an important element of quality control. Without validation there can be no assurance that the test results will be reliable and fit for the purpose. In some fields, validation of methods is a regulatory requirement. Generally, any method used to produce data in support of regulatory (e.g., FDA) filings or the manufacture of devices for human use should be validated.

All are candidates for validation, though the process for each can vary. Most test methods exist as validated standards, methods developed by technical standard organizations (ANSI, ASTM, ISO...) to establish uniform methods and procedures for testing. But standard methods do not always fit the requirements of the tests to be performed.

When should methods be validated?

The risk associated with products an output (dimensions, features, chemical requirements etc.) often dictates the validation activity based on the potential level of patient harm weighed against the business risk of not performing the activities.

The device risk index or harm classification dictates the minimum level of statistical confidence required. Higher risk requires more rigorous testing and higher levels of statistical confidence. In most cases, TMV is not mandated in the medical device industry (except ISO 11607). But demonstrating the safety and effectiveness of a device is difficult to do if the methods for establishing these parameters are not shown to be appropriate and reliable. Test Method Validation may be required for:

> ➢ A new method is developed
> ➢ A revision to established methods
> ➢ Methods that are moved or transferred

Changes Requiring Re-Validation

Take the case where a standard test method is established and in operation. However, a change to the system software is required. This type of change could impact the measured output. Therefore the change needs to be considered for re-validation.

Any other changes to the test procedure such as a change in handling of test specimens or the change, addition, removal or modification of equipment including fixturing can impact the measured output.

It is important to note that validation of a test method is not required on each individual piece of equipment or fixturing, once replicate equipment or fixturing is assessed during the validation study.

Some examples of changes not necessarily requiring any re-validation or change to a validation report etc. include:

➢ Clerical corrections to the test method that do not change the method or affect the measurement of the output.

➢ Removing of referenced supplies that do not impact test output, for example lint or cleaning agents.

➢ Movement of equipment does normally not merit re-validation of the test method, but a limited equipment qualification may be required.

Test method validations should be product and site specific. This means the site and product should be clearly defined in the scope of the test method validation documents. Before an already existing validated test method can be used with a new product or at a new site, the suitability of the existing test method must be documented. Suitability reports are examined further along in this book.

Regulatory Overview

US Food and Drug Administration

The FDA defines a medical device as *"A medical device is an instrument, apparatus, implement, machine, contrivance, implant or other similar or related article, including a component part, or accessory which is;*

Recognized in the official National Formulary, or the United States Pharmacopoeia, or any supplement to them,

Intended for use in the cure, mitigation, treatment of disease, in man or other animals, or

Intended to affect the structure or any function of the body of man or other animals, and which does not achieve any of it's primary intended purposes through chemical action within or on the body of man or other animals and which is not dependent upon being metabolized for the achievement of any of its primary intended purposes."

The Code of Federal Regulations (CFR) Title 21 Part 820: Quality System Regulation (QSR) 21 does not clearly call out requirements on method validation. It does not actually mentions the words "method" and "validation" side by side. However, many warning letters have been issued to manufacturing on the subject since at least 2005. Method validation also protects the manufacturer from allowing defective product to circumvent inspection methods if not fit for purpose.

Regulating guidelines from a variety of sources covered in the sections below. The discussion, as it relates to method validation, is somewhat circuitous for medical devices. As stated previously, this is caused by the absence of the phrase method validation in FDA QSR documents. For this reason, some basic treatment relating to process validation is covered below even though this topic is covered in detail in a separate chapter especially because the actual CFR definitions are general enough to lump methods into the category of a process if a CDRH auditor sees fit to do so.

The first sentence of the 21 CFR 820 Quality System Regulation scope states:

*"Current good manufacturing practice (cGMP) requirements are set forth in this quality system regulation. The requirements in this part govern the **methods** used in, and the facilities and controls used for, the design, manufacture, packaging, labeling, storage, installation, and servicing of all finished devices intended for human use."*

In this first sentence, FDA has deemed the topic of methods as not excluded from the purview of FDA. This interpretation is evidenced by the warning letters and Form 483's issued to Medical Device companies described in later sections of this chapter.

FDA Title 21 Code of Federal Regulations Part 820.3

*(z) **Validation** means confirmation by examination and provision of objective evidence that the particular requirements for a specific intended use can be consistently fulfilled.*

(1) **Process validation** *means establishing by objective evidence that a process consistently produces a result or product meeting its predetermined specifications.*

(2) **Design validation** *means establishing by objective evidence that device specifications conform with user needs and intended use(s).*

(3) **Verification** *means confirmation by examination and provision of objective evidence that specified requirements have been fulfilled.*

(g) **Design validation**. *Each manufacturer shall establish and maintain procedures for validating the device design. Design validation shall be performed under defined operating conditions on initial production units, lots, or batches, or their equivalents. Design validation shall ensure that devices conform to defined user needs and intended uses and shall include testing of production units under actual or simulated use conditions. Design validation shall include software validation and risk analysis, where appropriate. The results of the design validation, including identification of the design, method(s), the date, and the individual(s) performing the validation, shall be documented in the DHF.*

(i) **Design changes**. *Each manufacturer shall establish and maintain procedures for the identification, documentation, validation or where appropriate verification, review, and approval of design changes before their implementation.*

(i) **Automated processes**. *When computers or automated data processing systems are used as part of production or the quality system, the manufacturer shall validate computer software for its intended use according to an established protocol. All software changes shall be validated before approval and issuance. These validation activities and results shall be documented.*

Sec. 820.75 Process validation

(a) Where the results of a process cannot be fully verified by subsequent inspection and test, the process shall be validated with a high degree of assurance and approved according to established procedures. The validation activities and results, including the date and signature of the individual(s) approving the validation and where appropriate the major equipment validated, shall be documented.

(b) Each manufacturer shall establish and maintain procedures for monitoring and control of process parameters for validated processes to ensure that the specified requirements continue to be met.

(1) Each manufacturer shall ensure that validated processes are performed by qualified individual(s).

(2) For validated processes, the monitoring and control methods and data, the date performed, and, where appropriate, the individual(s) performing the process or the major equipment used shall be documented.

(c) When changes or process deviations occur, the manufacturer shall review and evaluate the process and perform revalidation where appropriate. These activities shall be documented.

<u>*Title 21 Code of Federal Regulations Part 820.80*</u>

(a) General. Each manufacturer shall establish and maintain procedures for acceptance activities. Acceptance activities include inspections, tests, or other verification activities.

(b) Receiving acceptance activities. Each manufacturer shall establish and maintain procedures for acceptance of incoming product. Incoming product shall be inspected, tested, or otherwise verified as conforming to specified requirements. Acceptance or rejection shall be documented.

(c) In-process acceptance activities. Each manufacturer shall establish and maintain acceptance procedures, where appropriate, to ensure that specified requirements for in-process product are met. Such procedures shall ensure that in-process product is controlled until the required inspection and tests or other verification activities have been completed, or necessary approvals are received, and are documented.

Title 21 Code of Federal Regulations Part 211.165 (e)

(e) The accuracy, sensitivity, specificity, and reproducibility of test methods employed by the firm shall be established and documented. Such validation and documentation may be accomplished in accordance with 211.194(a)(2).

Title 21 Code Federal Regulation Part 210.3 (b)

(b) The following definitions of terms apply to this part and to parts 211, 225, and 226 of this chapter.

(1)Act means the Federal Food, Drug, and Cosmetic Act, as amended (21 U.S.C. 301et seq.).

(2)Batch means a specific quantity of a drug or other material that is intended to have uniform character and quality, within specified limits, and is produced according to a single manufacturing order during the same cycle of manufacture.

FDA requirements for medical devices, for validation apply to:

Processes—PPQs for processes including but not limited to manufacturing, sterilization, cleaning and packaging

Software—Any used in the device itself, manufacturing and test equipment, process control or quality processes

Equipment—IQ/OQ/PQs for Test Equipment and Manufacturing Equipment

The Quality System (QS) regulation defines process validation as establishing by objective evidence that a process consistently produces a result or product meeting its predetermined specifications [820.3(z)(1)].

The requirement for process validation appears in section 820.75 of the Quality System (QS) regulation. The goal of a quality system is to consistently produce products that are fit for their intended use. Process validation is a key element in assuring that these principles and goals are met.

Processes are developed according to the design controls in 820.30 and validated according to 820.75. The process specifications, hereafter called parameters, are derived from the specifications for the device, component or other entity to be produced by the process.

The parameters are documented in the device master record per 820.30, 820.40 and 820.181. The process is developed such that the required parameters are achieved. To ensure that the output of the process will consistently meet the required parameters during routine production, the process is validated.

The basic principles for validation may be stated as follows:

- *Establish that the process equipment has the capability of operating within required parameters;*

- *Demonstrate that controlling, monitoring, and/or measuring equipment and instrumentation are capable of operating within the parameters prescribed for the process equipment;*

- *Perform replicate cycles (runs) representing the required operational range of the equipment to demonstrate that the processes have been operated within the prescribed parameters for the process and that the output or product consistently meets predetermined specifications for quality and function; and*

- *Monitor the validated process during routine operation. As needed, requalify and recertify the equipment.*

Per CDRH, the additional validation related definitions are:

Installation qualification: establishing documented evidence that process equipment and ancillary systems are capable of consistently operating within established limits and tolerances.

Process performance qualification: establishing documented evidence that the process is effective and reproducible.

Product performance qualification: establishing documented evidence through appropriate testing that the finished product produced by a specified process(es) meets all release requirements for functionality and safety.

Where process results cannot be fully verified during routine production by inspection and test, the process must be validated according to established procedures [820.75(a)]. When any of the conditions listed below exist, process validation is the only practical means for assuring that processes will consistently produce devices that meet their predetermined specifications:

Method validation as a requirement is not called out specifically; the FDA has issued warning letters and 483s in relation to the lack of "Method Validation".

W.H.O. Guidance

The World Health Organisation issued draft guidance for Test Method Validation of in vitro diagnostic medical devices in December 2016. Technical Guidance Series (TGS) for WHO:

Guidance on Test Method Validation of In Vitro diagnostic medical devices TGS-4

The guidance defines the terms verification and validation as follows:

*"**Verification** is the documentary proof that particular specifications have been met. When designing and developing an IVD, relevant attributes such as cost, and those for performance such as precision, sensitivity and stability are identified and given numerical specifications in design input documentation. It is subsequently the role of the R&D department to design an IVD that will meet those specifications.*

The R&D department consequently identifies valid test methods to demonstrate that the specifications have been met (verification) in the new design. Once design has been established, further numerical specifications are produced by the R&D department to ensure that the specifications of each attribute will be met consistently in routine production to ensure quality manufacturing. These new specifications are assigned to control critical production points and may include those for acceptance of raw materials, in-process materials, cleanliness of equipment, qualification of instrumentation and for the finalised IVD to verify its manufacture.

Again, it is also the role of the R&D department to identify appropriate test methods to monitor these specifications. An example of verification is related to incoming goods inspections; each time a raw material is purchased its properties will be verified against the specification using a validated test method."

Validation *is the documentary proof that the particular requirements for a specific intended use can be consistently fulfilled . Validation is defined as "verification against needs for a specific use" (i.e. the specification for that use). Within this guide, consistency is essential: it is an expectation that every lot of an IVD will behave as all other lots and will continue to meet design inputs. To ensure this, it is necessary to have validated test methods for measuring and/or monitoring specifications that will consistently produce results fit for purpose. The test methods must be validated to ensure that the results of measuring and/or monitoring are meaningful. For example, the need for accurate measurement of a raw material weighed in micrograms will not be achieved by using a weighing device with tolerance measured in grams. A test method using such an instrument would not be valid for the intended use. Thus, for the example provided, a test method should be specified that has the necessary accuracy and precision for measuring such weights, and an instrument and procedure identified that will consistently achieve this requirement during its use. The test method is then validated to produce results fit for purpose.*

Validation of a test method is distinct from its characterisation. Characterisation is documentation of some or all of the features of the method; validation is ensuring that the relevant characteristics are appropriate for the specific intended use. Validation of a method to be used widely, and for standard methods, often begins with complete characterisation. However, for each specific intended use it is likely that only a subset of the characteristics will be relevant and must be evaluated.

ISO 13485, Medical Devices Standard

Introduction

ISO 13485 is the quality management standard of choice for manufactures of medical devices. Revised in 2016, ISO 13485:2016 "specifies requirements for a quality management system where an organisation needs to demonstrate its ability to provide medical devices and related services that consistently meet customer and applicable regulatory requirements."1 The scope of the standard can apply to any organisation or company involved throughout the life-cycle of a product, including design and/or development, production, storage and distribution, installation, or servicing of a medical device and design and development or provision of technical or professional services.

The recent revision is designed to address recent developments in quality management and other updated regulations that relate to the industry. Improvements in the new version of the standard include broadening its applicability to include all organisations involved in the life cycle of the product, from the concept stage to end of life along with greater alignment with regulatory requirements and post-market surveillance including complaint handling.

ISO 13485:2016 is also used by suppliers or external vendors that provide QMS related management system- services. Requirements of ISO 13485:2016 are applicable to organisations regardless of their size and regardless of their type except where explicitly stated. Wherever requirements are specified as applying to medical devices, the requirements apply equally to associated services as supplied by the organisation. If any requirement in Clauses 6, 7 or 8 of ISO 13485:2016 is not applicable due to the activities undertaken by the organisation or the nature of the medical device for which the quality management system is applied, the organisation does not need to include such a requirement in its quality management system. For any clause that is determined to be not applicable, the organisation records the justification as part of their certification and quality management system.

Basic Definitions as defined in EU Annex IX of Directive 93/42/EEC)

Intended Purpose: Intended purpose means the use for which the device is intended according to the data supplied by the manufacturer on the labelling, in the instructions and/or in promotional materials. (Chapter I section 1 of Annex IX of Directive 93/42/EEC)

Transient: Normally intended for continuous use for less than 60 minutes.

Short Term: Normally intended for continuous use for not more than 30 days.

Long Term : Normally intended for continuous use for more than 30 days.

Invasive Devices: A device which, in whole or in part, penetrates inside the body, either through a body orifice or through the surface of the body.

Body Orifice: Any natural opening in the body, as well as the external surface of the eyeball, or any permanent artificial opening, such as a stoma.

Surgically Invasive Device: An invasive device which penetrates inside the body through the surface of the body, with the aid of or in the context of a surgical operation.

Implantable Device: Any device which is intended:

- to be totally introduced into the human body or,
- to replace an epithelial surface or the surface of the eye, by surgical intervention which is intended to remain in place after the procedure. Any device intended to be partially introduced into the human body through surgical intervention and intended to remain in place after the procedure for at least 30 days is also considered an implantable device.

Medical Device: means any instrument, apparatus, appliance, material or other article, whether used alone or in combination, together with any accessories or software for its proper functioning, intended by the manufacturer to be used for human beings in the:

- diagnosis, prevention, monitoring, treatment or alleviation of disease or injury.

- investigation, replacement or modification of the anatomy or of a physiological process.

- control of conception which does not achieve its principal intended action by pharmacological, chemical, immunological or metabolic means.
A medical device may be assisted in its function by the following means:

Active Medical Device: any medical device relying for its functioning on a source of electrical energy or any source of power other than that directly generated by the human body or gravity.

Active Implantable Medical Device: any active medical device which is intended to be totally or partially introduced, surgically or medically, into the human body or by medical intervention into a natural orifice, and which is intended to remain after the procedure.

Custom-Made Device: means any active implantable medical device specifically made in accordance with a medical specialist's written prescription which gives, under his responsibility, specific design characteristics and is intended to be used only for an individual named patient.

Device Intended for Clinical Investigation: any active implantable medical device intended for use by a specialist doctor when conducting investigations in an adequate human clinical environment.

Intended Purpose: means the use for which the medical device is intended and for which it is suited according to the data supplied by the manufacturer in the instructions.

Putting into Service: means making available to the medical profession for implantation.

Where an active implantable medical device is intended to administer a substance defined as a medicinal product within the meaning of Council Directive 65/65/EEC of 26 January 1965 on the approximation of provisions laid down by law, regulation or administrative action relating to proprietary medicinal products (6), as last amended by Directive 87/21/EEC (7), that substance shall be subject to the system of marketing authorisation provided for in that directive.

Where an active implantable medical device incorporates, as an integral part, a substance which, if used separately, may be considered to be a medicinal product within the meaning of Article 1 of Directive 65/65/EEC, that device must be evaluated and authorised in accordance with the provisions of this directive.

ISO 13485 & Regulations

In Europe, EN ISO 13485:2013 helps companies meet the requirements of: Directive 93/42/EEC on medical devices. This harmonised standard gives companies the "presumption of conformity" to complying with directives.

EN ISO 13485 was published in February 2013 and harmonised in August 2013 to cover the three directives:

- 90/385/ECC– The Active Implantable Medical Devices Directive (AIMDI)
- 93/42/ECC – The Medical Devices Directive (MDD)
- 98/79/EEC – In Vitro Diagnostic MDD (IVDMDD)

Overview of Standard

ISO 13485 has 8 Clauses or Sections which make up the structure of the standard.

Section 0 Normative References, Definitions and Terms

Section 1 Requirements of the Quality Management System (QMS)

Section 2 Normative References

Section 3 Terms and Definitions

Section 4 Requirements of the Quality Management System (QMS)

Section 5 Management Responsibility

Section 6 Resource Management

Section 7 Product Realisation

Section 8 Measurement, Analysis and Improvement

With regard to Test Method Validation, the relevant areas of ISO 13485 include:

(1) Clause 7: Product Realisation- Section 7.3 Design and Development

(2) Clause 8: Measurement Analysis

Clause 7: Product Realisation- Section 7.3 Design and Development:

Design and Development Verification and Validation ensure that the product is designed, developed and subsequently manufactured meeting all the customer requirements, regulatory requirements and business requirements. These requirements are classed as inputs to the design and development, and verification and validation ensure the inputs have been adequately taken into account.

The design and development testing sometimes replicate the commercial applications of the medical device, hence providing a realistic challenge in order to have confidence in the medical device.

Design Control

Design control is a necessary practice that ensures good engineering principles are maintained throughout the design phase of a product. It also refers to the continual design and development of the product through its very lifecycle. The design and development files and history must be controlled and maintained, with any changes properly assessed, tested and documented.

Clause 8: Measurement Analysis:

Clause 8 includes:

8.1 General requirements
8.2 Monitoring and measurement
8.3 Control of nonconforming products
8.4 Analysis of data
8.5 Improvement

8.1 General Requirements

Measurement, analysis and improvement are the key themes of clause 8. As with all medical devices, inspection and testing both during manufacturing and post manufacturing is necessary to ensure products and services function as intended and without defects. With any type of measurement or inspection analysis, the method used to complete the testing is critical. The method must be fit for purpose, and the equipment must be suitable. This "method validation" typically is done during the design and development phase.

8.2 Monitoring and Measurement

Monitoring and measurement are dependent on the information or feedback provided from various sources. The most important feedback is the post-production feedback that is gathered from customers or the end user. Again, this occurs over the whole lifetime of the product or service in question. There are a number of methods that can be used to obtain feedback. Some examples include:

-Customer surveys
-Customer complaints
-Review of regulatory databases such as MAUDE.
-Repair and servicing information

8.3 Control of Nonconforming Product

Non-conforming product presents a risk to patients or users of medical devices. When a situation arises where non-conforming product is manufactured or detected through inspection processes, the product must be controlled and segregated to prevent unintended use or distribution.
Some examples resulting in non-conformance are:

- When a manufacturing process drifts outside its validation window or operating parameters.
- A certificate of analysis for a raw material is not

provided by the supplier or the results are out of specification.

- In-process testing was not completed at the defined intervals.
- Training of personnel completing tests is not current or is inadequate.

8.4 Analysis of Data

In any engineering activity, data and the quality of the data is a key factor in making the right decisions. Provided the data collected is relevant and accurate, analysis of data can provide important insights into process performance, quality control and product functionality. Data should be collated in a consistent way and controlled by a procedure. When it comes to medical device manufacturing, the sources and types of data are multiple. Data can be generated from in-process testing and data can be generated from end of line testing aka finished product testing.

8.5 Improvement

ISO 13485 fosters a culture of continual improvement. As we have seen, each activity can be described as a process. For example, a manufacturing process, a procurement process, a complaints process. The set of processes that make up the quality management system need to be continually reviewed to ensure they are suitable and effective for the task at hand. Typical tools used to keep improvement in mind include:

- Review of the quality policy and quality objectives
- Frequency and category of corrective and preventative actions (CAPA's)
- Customer complaints
- Management review input

Definitions and Key Concepts

Attribute: is defined as the result of a property or characteristic. It is generally used with the terms pass or fail.

Accuracy: can also be defined as trueness. An expression of the closeness of agreement between the value that is accepted, either as a conventional true value or an accepted reference value and the value obtained. A system with low bias implies good accuracy and vice versa.

ANOVA (Analysis of Variance): a statistical method used to evaluate the significance of differences in means due to different factor-level combinations.

Bias: The difference between observed "average of measurements" and a reference value; also referred to as accuracy.

CQA (Critical-to-Quality): a property or characteristic with specific nominal value and appropriate limit and range providing a particular quality attribute.

Critical Process Parameter (CPP): a process parameter that has a direct impact on critical quality attributes.

Dichotomous Variable: an output with only two possible values. Also known as dummy or indicator variable.

Equipment Qualification: establishing documented evidence that the process equipment is suitable for the intended use and is capable of consistently operating within established limits and tolerances under normal operating conditions.

Process Validation: process validation is defined as confirmation via documented evidence that a particular process performs consistently to a high degree of assurance in accordance with predetermined specifications under anticipated conditions.

Measurement Capability Index (MCI): the Measurement Capability Index (MCI) represents the capability of the measurement system. It is used to evaluate the capability of the gauge to classify product against predetermined specifications.

MSA: a study to determine the degree of error involved in measuring the given parameter. The measurement system involves the combination of operations, procedures, gauges, instruments, environmental conditions, people and software.

Precision: the degree of agreement (scatter) between a series of measurements when a method is applied repeatedly to multiple samplings of a homogeneous sample or artificially prepared sample under the prescribed conditions. There are three types of precision; repeatability, intermediate precision and reproducibility.

Range: range is defined as the interval between the upper and lower measurements required. The minimum specified range should be within the equipment range and validated to operate at all points within the range.

Ruggedness (Intermediate Precision): variation on different days or with different analysts and equipment. The extent to which intermediate precision should be established depends on the circumstances under which the method is intended to be used.

Resolution: the smallest unit of measure that can be obtained reliably from a measurement device, also known as gauge discrimination.

Gauge R&R: represents the estimate of the measurement variation. The measurement variation has two components; repeatability or the precision under the same operating conditions (same operator, test method, sample, etc.) and reproducibility or the precision between operators when measuring the same sample with the same gauge.

Variable: is generally the output that is measured.

Validation: confirmation by examination and provision of objective evidence that the particular requirements for a specific intended use can be consistently fulfilled.

New Test Methods

A test method procedure should be created as early on as possible and trialed and examined for completeness and appropriateness.
If new test methods are required, a revision controlled draft should be available for the purposes of the test method validation.

Changes to Existing Methods

If changes to existing test methods are required, a redlined version highlighting the changes should be made available for the test method validation.

Method Transfer

If an existing test method is suitable for the test method validation, a suitability report can be completed to document the suitability and show that all factors have been considered (see attachment 1). However, the test method should have been previously validated. The parameters at which the validation is to be conducted must be within the existing validated range.

Equipment used in a test method must be assessed to ensure the process is within the equipment qualification. All validation testing must be done on qualified equipment. Equipment qualification is therefore a prerequisite of test method validation.

Test Method Ruggedness Study Protocols

Ruggedness refers to the variation, on different days or with different operators or equipment. The extent to which ruggedness (aka intermediate precision) should be established depends on the circumstances under which the method is intended to be used.

An initial ruggedness assessment should be completed to understand the sources of variation. More formal ruggedness studies may be required which should be captured in a formal study protocol.

The output of any ruggedness studies should detail any changes or modifications to the test method procedure.
Generally, a scoring system is used to describe ruggedness which forms a ruggedness assessment. As a result of ruggedness studies and consequent updates to the procedure, the ruggedness assessment needs to be reassessed. This reassessment should be reflected in the final scores of a Ruggedness Assessment Matrix.

Accuracy

Accuracy is a measure of exactness of the test method output or another way of putting it is the closeness of agreement between a set of test results.

For example, take a component that weighs exactly 4 kg according to an NIST traceable scale. If the weight of component is taken 10 times on the balance under study using the test method under study then calculate the mean weight of the 10 readings. The offset between the mean weight and the 4kg "accepted reference value" is a measure of bias.

A large bias = poor accuracy. A small bias = good accuracy. It is important to note that accuracy does not address the variation between individual measurements.

Simply put, if the average is very close to 4kg, then the test method could have been declared to be very accurate.

It is advised that you consult any relevant standards (e.g. ISO, ASTM) to the product or feature being measured as standards often will call out an accuracy requirement. Generally, results should be accurate to $\pm 1\%$ of the measured value. Therefore, the equipment must be fit for the intended purpose or the measurements in mind.

Note: instrument or equipment accuracy can normally be found on calibration certs provided by the manufacturer or vendor.

Precision

The precision of a method is the degree of agreement among individual test results when the same test method or procedure is applied repeatedly to multiple samplings that represent a population.

Precision can be a measure of either the degree of reproducibility or of repeatability of the method.
Repeatability refers to the use of a method using the same operator/test person with the same equipment. Repeatability should be assessed using either a minimum of 9 determinations covering the specified range for the method (e.g. 3 concentrations /3 replicates each). Reproducibility refers to the use of the analytical method in different laboratories such as in a collaborative study.

Ruggedness

Intermediate precision (also known as ruggedness) expresses differences related to laboratory variation, as on different days, or with different analysts or equipment within the same laboratory. The extent to which intermediate precision should be established depends on the circumstances under which the method is intended to be used. The effects of random events on the precision of the analytical method should be established. The use of experimental design (matrix) may be used to study the effects of typical variation (dominance factors) on the analytical method (e.g. equipment, analyst, days).

Representative/Continuous Sampling

Representative sampling is used to determine overall process performance (e.g. Pp / Ppk), which is more applicable for processes known or suspected as less than stable or not in statistical control. Sampling in this way best determines overall spread, which includes within-time and time-to-time variation.

Below, some examples are given on how to sample representatively:

1.   Sampling over a given time-period: e.g. a tray of product is produced every 15 minutes, the period of interest is a 1 hour interval and the sample size is 40.

2.   Sampling a batch or product lot not assembled in any order: if the product is packed in a tray (without any grouping) then sample from various sections of the tray.

Consecutive Sampling

This type of sampling involves taking one sample immediately after each another for the subgroup or time period in question, and is used to determine process capability (e.g. Cp / Cpk).

Consecutive sampling is used in particular to create control charts where a process is sampled in time order by selecting a subgroup sample consecutively and repeating this sampling over a number of subgroups while in same time order.
This method is typically used when the process is stable as there will be little or no causes of lot-to-lot variation.

Range

The range is defined as the interval between the upper and lower measurements required. The minimum specified range should be within the equipment range and validated to operate at all points within the range.

If an existing test method or piece of equipment is to be used, it is important to determine if the method parameters for the new/modified test method are within the validated range of the equipment qualification. Remember, all validation testing must be done on qualified equipment. Typically, the equipment qualification assessment is documented in the test method validation protocol.

Resolution

We have previously defined resolution as the smallest unit of measure that can be obtained reliably from a measurement device or system.

For example, a Vernier callipers may have different models with different resolutions. Some will have only two digits to the right of the decimal point (X.XX mm) and other models could read three digits to the right of the decimal point (X.XXX mm).

The instrument resolution should be better than the resolution of the product specification. If the product specification is X.XXX, then at least a "four-digit" measurement device should be used.

Probability Of False Alarms P (Fa)

This signifies the likelihood of rejecting a conforming unit. This is typically an acceptance criterion for attribute tests. Refer to MSA template for further illustration.

Probability Of Misses P (M)

This indicates the likelihood of accepting a non-conforming unit. This also is typically an acceptance criterion for attribute tests. Refer to MSA template for further illustration.

Validation Protocols

Typically, an approved template is used to create a validation protocol. The protocol sets out the approach to the validation i.e. the approach to qualify the test method. Refer to the appendix for an example of a validation protocol template.

Attachments to the protocol should include ruggedness assessments completed and references to supporting studies/reports. The drafted or "redlined" test method should be attached to the protocol also. The type of MSA protocol (attribute or variable) should also be determined in the validation protocol.

What Can Impact the Accuracy of a Test Method?

Accuracy is influenced by both the instrument (scale) and the test method. If you drop the object on the scale and take a reading before the scale has stabilised, the accuracy is likely to be poorer than when using a test method that demands allowing the scale to stabilise.
Examples include: Tensile strength at break - strength does not exist as a material property independent of the test method used to measure it.

For properties like time, distance, and mass, there are NIST traceable standards that can be measured. These standards have a generally accepted reference value that can be compared to the observed readings to assess accuracy (bias). No such reference sample exists for tensile strength at break, deflation time or implant radial strength. For tests without a reference value, the accuracy of the underlying sensor (e.g. load cell) used to determine the output should be addressed if possible.

MSA Studies

A measurement system analysis (MSA) is an experimental design used to identify the elements that affect measurement variation. There are two types of data in which MSA studies can be completed i.e. variable data and attribute data. These terms are defined below. Variable data: data that can assume a range of numerical responses on a continuous scale. Most measurements yield variable data.

Attribute data: data that represents the absence or presence of a characteristic.

Non-destructive tests: test where the measured characteristic is not altered due to testing. Since the sample is not altered, multiple readings can be taken on the sample with the expectation of getting the same measured result.

Destructive tests: test where the measured characteristic is changed due to testing. Since the sample is changed, there is no expectation of getting the same measured result over multiple readings.

So, in summary that makes up four types of MSA studies:

> ➢ Variable / Non-Destructive
> ➢ Variable / Destructive

- ➢ Attribute / Non-Destructive
- ➢ Attribute / Destructive Table

The following sections describe the requirements, measurement capability indexes and the typical acceptance criteria per MSA type.

General MSA Requirements

Test Environment Conditions - the test environment (i.e. temperature, humidity) should represent the conditions going forward. The effect of multiple environmental conditions can be evaluated if the study is properly designed and planned.

Sample Range - samples should cover the expected range of measurements.

Standard (for attribute MSA) - define the true answer (pass or fail). The standard is based on the inspection ratings of an expert opinion or a measurement system with known better inspection capability than the one under evaluation.

Measurement Instructions/ Training - follow the inspection instructions as defined in the controlled documents or redlines included with the protocol. Do not minimise variability by adding special instructions not defined in the controlled documents or redlines included with the protocol. Reference the controlled documents in the protocol. Special instructions are allowed when using pseudo samples provided that the variability is not minimised due to the instructions. Testers should have a high degree of skill and experience. Do not use new personnel or inexperienced people to conduct measurement studies.

Equipment Qualification and Calibration – The equipment must be calibrated prior to conducting the study. Evidence of the calibrated state should be documented in the report (e.g. calibration certificates etc.). It is important not to re-calibrate the equipment during the study as results can be different due to the calibration effect. The effect of calibration can only be evaluated if the study is properly designed.

Randomisation –

1. Assign the samples to the first operator in random order. Operator measures the parts.

2. Assign the samples to the second operator in random order. Operator measures the parts.

3. Assign the samples to the third operator in random order. Operator measures the parts. Repeat the process described in steps 1 to 3 with the operators for a second and third trial.

Data collection - when documenting the results of a trial, the operator should not have access to the results from the previous trials. A different data collection sheet must be provided for each operator involved in each trial. In lieu of a different data sheet, a data recorder may be used to blind the data recording operator to the test data of previous runs.

Variable MSA Studies

Non Destructive/Variable Msa Studies

The key requirements for non-destructive and variable MSA studies include:

No. of Operators – at a minimum, 3 operators should be used during the study. More operators are also recommended if human/operator interaction is a source of measurement error.

Sample Size – a minimum of 10 units is recommended.

Trials - a minimum of 3 trials should be completed.

Destructive/Variable Msa Studies

If a test is destructive in nature, repeated measurements cannot be taken as the sample is damaged or destroyed as part of the test. One solution is to adopt standardisation of units where homogeneous samples are created by standardising the material or manufacturing process.

- ➢ No. of Operators – 3
- ➢ Sample Size – 10 units
- ➢ Trials – 3 trials

This equates to 90 measurements in total. If standardisation is not feasible, the use of non-destructive pseudo-samples can be used. However, equivalence should be demonstrated between the pseudo sample and "true" units.

Attribute MSA Studies

Non destructive

The recommended and minimum sample size requirements for attribute/non-destructive MSA studies are shown below:

Recommended Minimum Sample Size Requirement

- ➢ # Operators - 3
- ➢ # Sample size - 25
- ➢ # Trials – 3

Destructive

When the test is destructive, repeated measurements cannot be taken as the sample is destroyed or altered. Some approaches are outlined below in order to quantify the measurement variability for destructive tests.

Standardisation Approach: homogeneous and representative samples are created by standardising the method of sample preparation, or material.

Sub-samples: cut each sample into three sub-samples to represent the three trials.

Pseudo-samples: create non-destructive pseudo-samples, documenting a rationale justifying the equivalence of the pseudo samples to the true samples.

Measurement Capability Index

The Measurement Capability Index (MCI) is calculated to assess the capability of the measurement system. The MCI is calculated as a % tolerance.

Measurement Capability Index acceptance criteria:

This index is used to evaluate the capability of the gauge to classify product against the specifications.

The index represents the % of the tolerance (upper specification limit (USL) and the lower specification limit (LSL) that is consumed by the measurement system variation. Figure 9 shows a graphical representation for this index.

Action Plan - Identifying Sources of Variation

Sources	Action
Measurement variation due to repeatability	Ensure the sample is not deformed over time due to repeated measurements and trails
	Ensure the equipment specification has the

	required precision
Measurement variation due to reproducibility	Review training and introduce standard work instructions

Suitability for Use Rationalisation Report Template

If an existing test method can be used with no or minor changes, a Test Method SFU Report can be used to document the test method validation.

Suitability for use report is appropriate only if the new product test method parameters are within the existing validated range.

If the test method parameters for the new product are outside of the validated range, the test method must be re-validated. Examples of cases which can utilise such suitability for use reports include:

> ➤ Test method transfer to a new manufacturing site.
> ➤ New product where the product specifications fall within the validated output range.
> ➤ Minor changes in component material which do not impact the validated test. Examples of changes that require full validation include:

New products:

> ➤ Extension of product sizes that fall outside the validated range.

Appendix 1

Suitability for Use Rationalisation
Report Template

Test Method Suitability for Use Rationalisation
Report

Product Name

Site

Author:

Date:

Approvals:

1.   PURPOSE

The purpose of this report is to document the justification for use of XYZ assuming that XYZ has been previously qualified and validated for product XYZ.

The conclusions of the test method validation/suitability for use are valid and can be leveraged for new product zzz to be tested at Site YYY.

2.   SUMMARY

This report documents evidence of the suitability for use of the test method XZY.

Table 1 - Variable Data

Test Method	Measurem ent	Precisio n	Validate d Range	Resolut ion
TMXX X	Tensile Strength	MCI = 16%	0% to 11%	0.01 %

Table 2 - Attribute

Test Method	Measure ment	Effectiv eness	P(FA)	P(M)
TM000	Laser Mark	99%	1%	1%

4. CONFORMANCE TO GUIDELINES AND STANDARDS

Reference	Title	Revision
EN ISOxyz	Xxyyzz	Date
FDA	Xxyyzz	Date

4. JUSTIFICATION

The following are the justifications for leveraging test method suitability for use Justification for Leveraging xxxyyyzzzz validation conclusions for new product xxyyzz at site xxyyzz.

Test Method	Leveraged Report(s)	Parameter	Leveraging Justification
X	XXYYZZ	Probability of False Alarms P(FA)	
		Probability of Misses P(M)	
		Effectiveness	

Test Method	Leveraged Report(s)	Parameter	Leveraging Justification
		Ruggedness	_____
Variable Outputs (Continuous Data)			
X	Xxxyyyzzz	Accuracy	The acceptance criteria for accuracy may be driven by a standard such as ISO, ASTM. Generally, the calibration documentation will detail the accuracy of test equipment. If the equipment is capable of measuring to 1% of the measured value, the system can be deemed accurate.

Test Method	Leveraged Report(s)	Parameter	Leveraging Justification
Y		Precision	MCI ≤30% is typically the acceptance criterion. MCI is calculated via data captured during an MSA (Measurement System Analysis). Using a statistical tool the MCI is calculated.
		Range	The test method range and the equipment range must both meet the requirements of the product feature to be measured.
		Resolution	Resolution is the smallest unit of measure that can be obtained reliably from a measurement device, also known as gauge discrimination. E.g. 0.01mm.

Appendix 2

Test Method Validation Protocol

Test Method Validation Protocol

**TEST NAME
PRODUCT/SITE**

Author:

Date:

Approvals:

1. PURPOSE

The purpose of this validation protocol is to document the requirements and acceptance criteria that will establish that test method XYZ is suitable for use with product X at site Y for testing the characteristics listed in Table 1.

Measurement	Specification
TEST 1-Tensile Strength	*XXXYYYYYYYY ZZZZZZZZ*
TEST 2- Dimensional Length	*XXXXXXYYYY YYYYYYYYYY YZZZZZZZZZZ ZZZ*

2.   SCOPE

The requirements and acceptance criteria as specified in the test method requirements.

3.   DEFINITIONS
For the purpose of this validation protocol, the following terms and definitions apply:

RPT   *Report (document prefix)*
TM   *Test Method (document prefix)*

5.   REFERENCES

Reference	Title	Revision
EN ISO XXXX	XYZ	DATE
FDA Guidance	XYZ	DATE

6.   BACKGROUND

Verification of Equipment Qualified Ranges

Equipment	Parameter	Qualified Range of Equipment	Test Method Range	Pass
XYZ	*Diameter*	*E.g. 1 to 20mm*	2.0mm – 15.0mm	*Yes*

7. CONFORMANCE TO GUIDELINES AND STANDARDS

Guideline/Standard	Measurement	Requirement
EN ISO XYZ		
FDA XYZ		

8. ACCEPTANCE CRITERIA AND RATIONALE

Paramet er	Requirem ent	Accept ance Criteri a	Verification Process
Accuracy	Referenc e accuracy requireme nts specified in standards . E.g. accuracy of ±2%	±2ºC	Review the instrument measurement accuracy by referencing calibration records

Precision	MCI.	MCI ≤ 30%	A Measurement System Accuracy GR&R Study to be completed.
Range	Define the range of measureme nts	Test method must be able to measur e values within the require d range.	Parameters and settings should be justified
Resolutio n	0.1mm	0.1 °C	The resolution of gauge or equipment as per calibration assessment.

Notes

ValidaitonResources.org